Solar and Wind Power:

Lessons How Buils Your Own Power Generating System At Home for Beginners

Disclaimer: All photos used in this book, including the cover photo were made available under a Attribution-ShareAlike 2.0 Generic (CC BY-SA 2.0) and sourced from Flickr

Copyright 2016 by Publisher – All rights reserved.

This document is geared towards providing exact and reliable information in regards to the topic and issue covered. The publication is sold with the idea that the publisher is not required to render accounting, officially permitted, or otherwise, qualified services. If advice is necessary, legal or professional, a practiced individual in the profession should be ordered.

- From a Declaration of Principles which was accepted and approved equally by a Committee of the American Bar Association and a Committee of Publishers and Associations.

In no way is it legal to reproduce, duplicate, or transmit any part of this document in either electronic means or in printed format. Recording of this publication is strictly prohibited and any storage of this document is not allowed unless with written permission from the publisher. All rights reserved.

The information provided herein is stated to be truthful and consistent, in that any liability, in terms of inattention or otherwise, by any usage or abuse of any policies, processes, or directions contained within is the solitary and utter responsibility of the recipient reader. Under no circumstances will any legal responsibility or blame be held against the publisher for any reparation, damages, or monetary loss due to the information herein, either directly or indirectly.

Respective authors own all copyrights not held by the publisher.

The information herein is offered for informational purposes solely, and is universal as so. The presentation of the information is without contract or any type of guarantee assurance.

The trademarks that are used are without any consent, and the publication of the trademark is without permission or backing by the trademark owner. All trademarks and brands within this book are for clarifying purposes only and are the owned by the owners themselves, not affiliated with this document.

Table of Contents

Solar and Wind Power: 1
    Lessons How Buils Your Own Power Generating System At Home for Beginners 1
Introduction: A Little Wind and Sun...4
Chapter 1: Going Solar..5
    A Little Solar Knowledge 5
    Solar Paneling..6
    Batteries And Solar Power...7
    Using an Inverter 7
Chapter 2: DIY Solar Power Generating Systems..8
    Basic Tools and Preparation 8
    Slightly Used and Worn Solar Cells 8
    Solar Panel Construction 9
    Installing Wiring 10
Chapter 3: Understanding the Wind 12
    Rating the Wind...12
    Understanding Wind Turbines 13
    Preparation for Wind Power Generating Systems 13
Chapter 4: Building Your Own Micro Wind Turbine 16
    Construct Your Rotor Blades 16
    Hooking Your Blades up to your Rotor 17
    Placing your Wind Turbine to Your Tower 18
    Batteries and Wind Power 19
    Understanding The Inverter..20
Chapter 5: Some Alternative Wind and Solar Power Projects.......................................21
    Soda Pop Solar Panels 21
    Construct a Windmill..22
    Solar Flashlight...23
    Mirror Frame Solar Paneling..24
Chapter 6: Pros and Cons of Solar and Wind Power..25
    Solar and Wind are Both Sustainable 25
    Wind and Solar are a Local Commodity 25
    Solar and Wind Power Are Clean 26
    Solar and Wind Power are Cost Effective 26
    Wind and Solar Power are Unpredictable 27
    Solar Panels and Wind Turbines Bother the Neighbors 27
    Wind and Solar Power Have Become Much more Affordable................................28
    Future Possibilities in Wind and Solar Energy 28
Conclusion: You Can't Beat the Sun and the Wind...30
FREE Bonus Reminder 31

Introduction: A Little Wind and Sun

Two things that you can be sure to count on as long as there is life on this planet; are the sun and the wind. In fact, the two are integrally connected with each other. Every single time the suns rays strike the Earth's atmosphere, the charged particles create the turbulence that we call wind. These solar particles travel on this wind and our aided by its force as the photons from the sun collect on countless solar panels strewn across the planet.

The fleeting wind can also be harnessed and captured in the spinning wind turbines that you may have seen planted across the countryside. In this book we are going to learn about the best aspects of solar and wind based energy systems. So get out there and get your constructive DIY juices flowing because a little bit of wind and sun can do you a whole lot of good!

Chapter 1: Going Solar

Using the sun for energy should come naturally to us, since all life on Earth depends on that powerhouse of nuclear fission floating in outer space. Technically the whole planet is solar powered already if you think about it. Flowers won't bloom, and birds won't sing, if the sun doesn't rise up above the horizon in the morning. In this chapter let's show our appreciation toward the sun by going completely solar!

A Little Solar Knowledge

Before we begin any project, solar or otherwise, we should know little bit more about the core concept behind it. And of course the core concept behind any solar powered device is the sun itself, so a little bit f solar knowledge would do you some good before you begin. The heat giving luminous object that we see shining in our sky that we call "the sun" is actually a star; it's a star no different than the billions of other stars you can see in the night sky.

The sun; our "day star", just appears larger to us because it is much closer than any other star in the universe. The sun is the closest star to the Earth, but it is still a daunting 93 million miles away from us. But even from this distance, just one second of the pure solar output that reaches the Earth has enough to power up your home all year long!

Unfortunately solar gathering technology is still in its infancy and some of the best solar apparatuses only manage to utilize about 10% of the energy the sun bombards us with, but even at 10% efficiency you will have more than enough solar power for your energy needs.

Solar Paneling

I hate to burst your bubble but solar panels don't actually absorb sunlight. The panels are just the protective case; it is the silicon cells inside the paneling that actually has the ability to absorb solar energy. But since "solar panels" are of such common parlance, in this book we will often default on the term "solar panel" as well. But yes, it is the solar cells that do all the work of solar absorption through silicon alloys that are able to absorb molecular photons of solar energy. These cells then transform these molecules into direct solar power.

Batteries And Solar Power

You will need some proper battery power in which to store and utilize your newfound solar energy. Standard deep cycle sealed batteries should do the trick just find. Try to buy your battery that you will use for this project early on, so that you won't be delayed!

Using an Inverter

Inverter's are crucial because they are the device that can convert a DC power current from your solar or wind powered system and then directly redirect it toward your electronic needs of AC power.

Chapter 2: DIY Solar Power Generating Systems

After discussing the principles and concepts behind solar power, lets get to work on making that solar energy become a reality for us. This chapter highlights some of the best solar power generation systems.

Basic Tools and Preparation

Most of the basic tools involved in the installation of a solar power system can be found at your local hardware store. You will need simple items such as glue, tabbing wire, wire cutters, and perhaps a pegboard to serve as your solar panel. You will also need to find a way to obtain solar cells.

Slightly Used and Worn Solar Cells

The great thing about solar cells is that you don't have to buy them brand knew. That's right! Even if they have been used before and have worn and sometimes even damaged surfaces, these cells will still work towards absorbing solar power! Even with cracks in their skin, solar cells will still absorb solar energy, so if you can obtain used or factory defective solar cells at a discount price your wallet will thank you later!

You just have to know where to get them. One of the best places to look for these slightly used solar cells is online. Craig's List is a great place to find local people in your area that are willing to hawk their solar cells for a low price. You can also go to Amazon to find some good deals on blemished solar cells as well.

And when we say blemished we are talking—in many cases—about solar cells that may have suffered some sort of minor defect when they were made in the factory, such as hairline scratch or crack that makes them just little less than perfect. These solar cells may look roughed up but the still work. So always be on the look out for slightly used and worn solar cells.

Solar Panel Construction

After you have obtained your solar cells you can begin work on constructing your solar panel material that will house them. Your solar panel should have a base that consists of a square shaped piece of wood. I have found that peg board works quite well for this task. Try to find a piece of peg board that is at least 2/8 of an inch thick with a 1 inch lip at the bottom of it to better hold the solar cells in place.

On this peg board, use a pencil to sketch out square patterns in which you can place your solar cells. With your pattern sketched out on your peg board, begin placing your solar cells into the squares you have drawn. Fix them in place by gluing the backs of the solar cells to the pegboard panel. Due to the rather obvious circumstance of these cells being placed in some rather high solar temperatures, make sure that the glue you use is heat resistant so that the glue bond doesn't break in the heat.

Give your glued on solar panels a few hours to dry, just to make sure that they are secure. With your solar cells in place take the peg board that they are glued onto and place it inside a wooden frame. I typically take two 2x4 pieces of wood, cut them each in half lengthwise and use them as framing for the peg board assembly.

Installing Wiring

With your solar cells securely inside your paneling, its time to think about how you are going to install your wiring. You electrical wires inside your solar panel, do not have to be incredibly complex, you can get the job done with just one simple drilled opening and a small length of wire. Just take a hand-drill and use it to create a small hole in the side of your panel.

After you have done this take your piece of electrical wiring and run it through the hole. Just drop his wiring down into the hole and use the other end so that you can connect it to a "blocking diode". This will enable your batteries to block energy from being used during nonuse of the battery.

This diode is attached directly to the wires protruding out of the paneling. After you have done this take your hand-drill and create another hole in which you can pull your wiring through and attach a polarizing, two-pin plug to the end of your wire and connect it to your battery. Now your DIY solar power generating system is complete!

Chapter 3: Understanding the Wind

The wind is a result of solar energy stirring up turbulence in our planet's atmosphere. In this chapter we will better learn how it is that this dynamic atmospheric disturbance that we call wind can be used to create wind energy for you and your household.

Rating the Wind

Just by going out on any given day of the week you can no doubt tell that the speed and impact of the wind changes from day to day. This is the reason we have meteorologists, so they can predict what the wind and other aspects of our weather may do on a daily basis. But weather forecasts aside, the easiest way to know what kind of wind speed you might be facing, is to rate it on a scale. It is for this reason that the "Beaufort Scale" was created so that you can classify wind on a scale anywhere from 0 to 12.

Knowing where the wind falls on this scale is crucial for the successful utilization of wind energy. For best results, you should have wind that hits right around 5 or 6 on the scale, anything beyond 6 will be too turbulent to maintain and most winds below 5 will not have enough power to sustain your generation of wind power. Rate the wind carefully so that your wind power system never runs out of steam!

Understanding Wind Turbines

Have you ever used an electric house fan on a hot summer day? You know, the kind you plug into the wall, and the blades spin in order to send cool air to your sweaty face? Well, amazingly, a wind turbine works in the complete reverse order. Instead of plugging a fan into the wall so an electric spark can cause blades to spin in order to create artificial wind.

A wind turbine uses naturally wind streams to spin it's blades, these spinning blades then move a shaft and send kinetic energy from the motion to a generator which will convert this kinetic energy into electricity. That's right, instead of using electric power to create wind, the wind turbine uses wind to create electric power!

Preparation for Wind Power Generating Systems

Besides the basic mechanics of the wind and how a wind turbine utilizes it, there are a few more preparations that you need to understand before you build your own wind power generating system. And the most important of these preparations is no doubt to figure how much wind your geographic location can generate in the first place.

Because after everything is said and done, the sheer volume of wind power that you can create will depend on the location you are in. Having that said, most places in the United States, Canada, and Northern Europe should have an ample amount of wind generating capacity. Regardless of these trends however, you should prepare yourself for just how much wind you can expect in the region in which you live.

And in order to succeed in your project you should live in a place that receives regular wind gusts of at least 7 miles an hour. Proper prep for a wind turbine also necessitates a plot of open land. And while you don't have to have a large farm in order to use a wind turbine, you will need at least half an acre of space in order to comfortably generate enough wind.

Along with these external measures of preparation you should also work to prep your home itself in order to make it as energy efficient as possible. And one of the number one ways we may expend energy, especially during the cold winter months, is through heating our house. So it goes without saying that you could save a lot on energy expenditure if you could just prep your house to retain as much heat as possible in the first place.

And the best way to do this is to make sure that the walls of your home are properly insulated. Make sure that your home's insulation is up to par, and no drafts are blowing through your home's structure so that you can save energy when it comes to heating. With your heating under control you should then move on to streamline your home's lighting.

Switching your standard bulbs with fluorescent light bulbs could do quite a bit when it comes to saving on your electric bill. These bulbs have been known to cut electric bills in half, and when it comes to reducing the amount of your precious wind energy that your home consumes, fluorescent bulbs send out softer less energy consume illumination that can save you money and energy. You should also take an inventory of the rest of your home appliances, taking note of what electric devices are expending the most energy and work to limit their expenditure.

Chapter 4: Building Your Own Micro Wind Turbine

With the basic concepts of wind power behind use, the time has now come to construct one of our own. There are a few ways to build your own micro wind turbine. But for the sake of this book, we will focus on one of the cheapest yet most practical methods of construction, showing you how to construct a wind turbine from the ground up. So let's get started!

Construct Your Rotor Blades

It is the blades of your turbine that will come into contact with the wind and convert the motion that it provides into electricity, so you need to make sure that you get this important component right. And one of the most important aspects of getting this right is to master something known as the "tip speed ratio" or "TSR". The TSR formula will show you how long your blades need to be in relation to wind speed.

Most important in this, is the tapering of the termination points of your blades. These dimensions should match with your general TSR protocol. In doing this you must build your blades out of wood that can withstand the wind speeds that it will be subjected to. But besides wood, it is also possible to construct rotor blades out of PVC pipe.

In order to carve your blades just take out a small hand-knife or other carving utensil and use it to begin whittling your blae down from its mid section all the way to the tip, making the blade narrower and narrower as you go. Once the appropriate shape of your blae has been fleshed out take some sandpaper and use it to further refine the material. Your rotor blades are now ready for use.

Hooking Your Blades up to your Rotor

With the creation of your blades you can then start the process of hooking them up to your rotor. In order for this installation to be successful your rotor needs to be triangular in shape so that you can put the blades in place properly. You will need to use a protractor to make sure that your angles are precise. Once this is established you can then drill a small hole right in the center of the rotor, so that you can fasten your blades in place. This is how you hook your blades up to your rotor.

Placing your Wind Turbine to Your Tower

So that your blades can do their best you need to place them up on top of a high tower so that they can access strong wind currents. Just by the laws of nature, you have to realize that the higher you place your wind turbine the better and stronger the gusts of wind will be. Having that said however, you do not want your turbines placed so high that it faces forces that are so turbulent that it rips the wind turbine from its moorings.

The height of your tower depends on the amount of wind in your area, but as a rule, in most flat land areas you should have a tower that reaches up at least 7 feet in the air. If the wind speed tend to greatly fluctuates in your area, you may want to consider investing in an adjustable tower that you can raise and lower at will. These are called "tilt up towers" and they are incredibly useful in the face of changing weather patterns. If the wind is too strong you can lower your tower, and if the wind is too light, you can raise it. Keep all of these things in mind when you consider the placement of your wind turbine on your tower.

Batteries and Wind Power

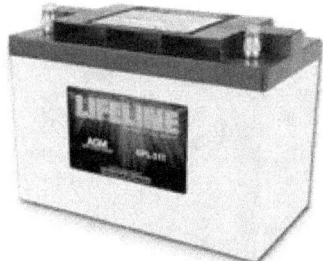

After creating wind energy you have to store it somewhere, and batteries are essentially storage units, that can be used to retain this power. The best battery to use for this purpose is what is known as a "deep cycle" battery. These kinds of batteries are either sealed or flooded. For a battery to be sealed it means that the battery cell inside of it is sealed off so that none of the chemicals inside can escape.

Completely opposite to this, is a flooded battery that has an open battery cell in which the battery's chemicals can openly flood out into the rest of the battery from the cell. A word to the wise on this one; if you live in the cold north, don't use a flooded deep cycle battery. Because these flooded batteries will no doubt have their chemical components freeze up in the declining temperatures. Even though flooded batteries are often cheaper, it is for this reason alone that sealed batteries are always preferable.

Understanding The Inverter

Just like with solar power, any good wind power system will need an inverter to change the direct current of your wind energy into an alternating current of usable AC power. Let your inverter take your wind channeled energy and convet it into direct power for your electronics.

Chapter 5: Some Alternative Wind and Solar Power Projects

After covering some of the more basic aspects of wind and solar power project construction, let's take a look at some of the bit more alternative wind and solar power projects you could come by.

Soda Pop Solar Panels

If you like sugary drinks, then this alternative DIY is for you! Because yes, believe it or not, the aluminum from soda pop cans works as a perfectly conducive material for soda pop cans. And you won't need that many either, you could do this project with as little as 30 cans. That's basically a 24 pack of soda plus a few extras, most soda drinkers can come up with this figure in just a couple of weeks, so it shouldn't be too much trouble.

But even if you don't care for drinking soda, you could probably choose any weekend of the month and gather up that many cans from the food court at your local shopping mall. You just might want to inform the mall security guards first so they don't confuse your aluminum can pillaging for shop lifting or some other nefarious deed. Regardless of how you obtain your cans however, once you get them, do as much as you can to clean them.

This is partially due to sanitary reasons, but it is also due to the simple fact that cans covered with the syrupy residue of sugary soda will face interference with their capacity to absorb solar energy. So rinse them off, scrub them, do whatever you can do to make those bad boys clean! After cleaning out your cans the next thing you are going to want to do is take a pair of scissors (or wire cutters) and cut through the middle of the cans lengthwise.

Once you have done this you should be able to fold the can out into a flat square. You might want to use a hammer and with the can spread out on a flat surface pound it evenly until it is as flat as possible. With your cans flattened, you can then begin laying them down inside your solar paneling.

For your solar paneling you just need a flat wooden surface, I usually use peg board for this task. Glue these flat cans onto your peg board keeping them close together in rows. Next you can drill a hole into the peg board and run your electrical wiring to the cans on one end and an AC adapter on the other. Your soda pop solar panels are now ready to go!

Construct a Windmill

The windmill stands as a classic testament to medieval engineering, but it shouldn't be relegated just to the middle ages. Because the truth is, this wind power generator can still be used quite successfully today. All you need for this alternative wind power DIY is a set of wooden dowels, some PVC pipe, a drill, and duct tape.

First take two 3 foot pieces of PVC pipe and put one on top of the other, gluing each to each. These combined pipes will serve as the base of your windmill. Next, attach your rotor blades to the top of this base by drilling a ole through the center of the rotors and tower head. Use a screw to fasten this assembly in place. Now start up an 80 degree angle using 16 inch pipe. And finally, put your motor on the back of the mill and your windmill is ready for use.

Solar Flashlight

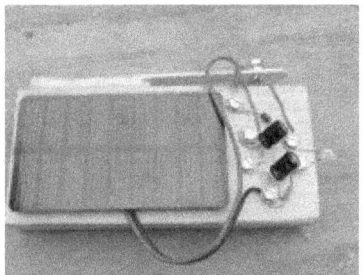

You never know when the power might go out and you will need a flashlight. The standard flashlight won't last long in an extended blackout however, so let's figure out how it is that we can have a solar powered flashlight. In order to get flashlight like this to work just get your hands on one solitary solar cell and insert it into a conventional flashlight.

You can tear out the battery carrying compartment and pop your solar cell inside. Once inside take the wiring that was previously attached to the flashlight bulb and the battery compartment and reroute it to your solar cell. Solder this wiring to your solar cell and you can now use direct solar energy to power this flashlight.

Mirror Frame Solar Paneling

The makeup of this project is fairly straight forward. Just take an old mirror and fill up its surface with solar cells. It turns out the mirror itself can work as a perfect platform for solar cells. Just glue your solar cells directly to the mirror's surface. With these solar cells in place you can the drill a hole in the side of the frame and insert a USB port right in the side. Properly wire up this USB to the solar collecting cells and you will have a solar power system that you can directly plug your phone and lap top into!

Chapter 6: Pros and Cons of Solar and Wind Power

Finally, at the end of this book I wanted to take the time to compare and contrast both the pros and the cons of solar and wind power.

Solar and Wind are Both Sustainable

This one should be the most obvious, but it is worth repeating again, both solar and wind power are sustainable for the foreseeable future. As long as the sun shines through our atmosphere the wind is going to blow, and as long as that sun shines we can also absorb that energy to benefit our lives. As long as our solar neighbor, the sun, doesn't get sucked up in a black hole, solar and wind power, as direct byproducts are both completely sustainable for the long haul.

Wind and Solar are a Local Commodity

Yes, it's true. The wind and the sun don't have to be shipped from Mid East Oil Wells, wind and solar energy can be extracted from right where you live! You will never see a country launch an invasion to seize the wind and sun! These commodities are naturally available for anyone. Wind and solar energy are a local commodity for anyone anywhere.

Solar and Wind Power Are Clean

The wind and sun rays floating through the Earth's atmosphere are clean. Instead of the pollution that you get from coal and oil, wind turbines and solar panels just take what is naturally already there in the environment. There is no pollution or harmful byproduct involved in the process whatsoever. These two clean energy systems do not add or take away; they only use what is already there. Solar and wind power are most definitely clean.

Solar and Wind Power are Cost Effective

Both wind and solar energy are very cheap to establish. And after your systems are up, you won't have to pay much of anything at all. Your wind and solar power system will actually begin to pay you back at a 5 cent to kilowatt level.

Wind and Solar Power are Unpredictable

Wind and solar ban both be unpredictable. Yes, some days are cloudy with no sun, and it might also become increasingly difficult to get a good wind. By its very nature wind and solar energy can be a bit unpredictable. But the best way to counter this unpredictability is to create a back up plan. This is where generators come in. Because if your system fails during the apocalypse, one simple switch to the generator can get you through the storm.

Solar Panels and Wind Turbines Bother the Neighbors

It's usually a given no matter where we live that we need to get along with our neighbors. But having a 20 foot wind turbine tower and your entire roof covered with solar panels, may be frustrating to some who are living near you. For a large project like this you may want to take note of it.

Wind and Solar Power Have Become Much more Affordable

Compared to the enormous costs that were once associated with setting up solar panels and wind turbines just 20 or 30 years ago, and the next to nothing amount involved with their construction today, both of these power systems are much more affordable. The price to maintain these systems have dropped by an astonishing 75% since 1985! These apparatuses have become rather frugal in their maintenance, and can now be created by even the cheapest of DIY enthusiast!

Future Possibilities in Wind and Solar Energy

Let's face it folks. Fossil fuels such as oil and coal are as dead as the dead animal material that gave them life. And as the oil and coal deposits continue to be depleted dry, there is nowhere else we can go but towards the future possibilities of wind and solar energy. Wind alone can provide an incredible 400 terawatts if it were fully utilized around the world, and the wattage of solar power, would be even more remarkable. The future really does depend upon wind and solar energy.

Conclusion: You Can't Beat the Sun and the Wind

No matter where you go or what you do, the sun and the wind will be following you. They lurk just behind your footsteps. With such a great natural resource so close to us, we never have to do without. No matter what anyone may say, you ust can't beat the sun and the wind. Thank you for reading!

FREE Bonus Reminder

If you have not grabbed it yet, please go ahead and download your special bonus report *"DIY Projects. 13 Useful & Easy To Make DIY Projects To Save Money & Improve Your Home!"*
Simply Click the Button Below

OR **Go to This Page**
http://diyhomecraft.com/free

BONUS #2: More Free & Discounted Books or Products
Do you want to receive more Free/Discounted Books or Products?
We have a mailing list where we send out our new Books or Products when they go free or with a discount on Amazon. Click on the link below to sign up for Free & Discount Book & Product Promotions.
=> **Sign Up for Free & Discount Book & Product Promotions** <=

OR Go to this URL
http://zbit.ly/1WBb1Ek

www.ingramcontent.com/pod-product-compliance
Lightning Source LLC
Chambersburg PA
CBHW030106230526
45471CB00003B/1284